"Close your eyes," said Mr. Prentice. "Now imagine that you could be anyone or anything at all. What would you like to be? Think about it, and then draw a picture or write your ideas. We'll post all the papers outside the office."

Madeline has no trouble thinking of what she would like to be. She imagines herself as a vet. She'll have a small office, where she'll examine people's pets. She would give them medicine to help them get well. Madeline can imagine nothing better than that.

Henry also has no problem with this project. He has always helped his father fix machines. Henry knows how to diagnose engine problems, such as dirty gasoline. Henry draws a picture to show himself working on an engine.

Leo has a different thought. He can imagine himself making videos of lions. Lions are strong, mobile animals that move across a large area. And Leo knows that his name has a special connection with the word *lion*.

Alice loves to imagine herself as the main character in a story she has heard or read. "I know!" thought Alice. "I'll create a picture of me as Alice in Wonderland! And I'll be having a wonderful adventure!" Alice is overjoyed with this idea and soon begins to draw.

Gene thinks of a great idea! He is learning to play the violin, so he practices every day. Gene imagines playing the violin in a concert. People come from all around the world to hear him. When he finishes, they are quiet for a moment. Then they burst into thunderous clapping.

Violet often likes to create poems in her mind. She gets ideas for her poems everywhere she goes. Violet can imagine herself as a famous poet. She will write poems, publish them, and read them aloud to children—in person and on TV!

When all the children are done, they share their papers with the class. Some papers have just pictures, some have just writing, and some have both. But all the papers are special. They show what is possible when you imagine. Everyone who goes to the office stops to read the papers. They get people to imagine what they themselves would like to be.